The Search for Extraterrestrial Life

Patrick H. Stakem

(c) 2019, 2022

Number 28 in the Space Series

Table of Contents

Introduction..4

Author...6

Life..8

Earth Life...9

Other Life in our Solar System?...13

    Mercury..14

    Venus..16

    Earth...17

    Near Earth Objects..17

    The Asteroid Belt, dwarf planets, and Centaurs.............18

    Comets and Meteorites..21

    Visitor from afar..22

    Mars...23

    The Gas Giants and their moons....................................27

        Jupiter...28

        Saturn..32

    The Icy Giants and their moons......................................35

        Uranus...35

        Neptune...37

    Pluto, and beyond...38

    NASA's Ocean Worlds Exploration Program..................38

The Search for Extra-terrestrial Life beyond our Solar system.....40

Panspermia...44

Abiogenesis..45

The Drake Equation..45

SETI...47

    Search in our Solar System 49

    Private Search...51

ExoPlanets..51

    Search for Exoplanets..52

The Drake Equation..53

    NEXSS...55

    Habitable Exoplanet Imaging Mission............................55

        Ross-128b...56

Alpha Centauri's planets...56
  Kepler-443b..57
  WASP-121b..57
Afterword..57
Glossary of Terms..58
References...64
Resources..71
If you enjoyed this book, you might also be interested in some of these...73

"In an infinite Universe, there must be other life. There is no bigger question." Stephen Hawking.

Introduction

The book discusses the topic of the search for Extraterrestrial life, that is, life not on Earth. The search is taking place in the planets and moons of our solar system, and on Exoplanets, elsewhere in our galaxy. The problem is, we only have 1, albeit vastly diversified, example of life. We can conjecture that life only has a chance in certain thermal ranges (the Goldilocks zone), so that's where we look. Unfortunately, strange life was observed at deep ocean thermal vents, with chemicals, pressures, and temperatures that would kill us in an instance. Are we a good example of "life"? Maybe not.

It would be terrible vain to suggest we are the only lifeforms in the Universe. The Universe is large and getting larger. We are the only forms of life we know. But, we haven't looked everywhere. Interestingly, most of the Earth's religions do not preclude extra-terrestrial life. It broadens our view of the magnificence of the universe.

The "easy" case is looking for other life in our solar system to expand our definition of "life. Then we are looking for bio-compatible planets, that we might use as a back-up.

We know about exo-planets, orbiting other stars, as well as exo-moons, and exo-solar systems. These are all very far away. We only have one known form of life that we know of. There is a whole lot of variation, but it is all

DNA-based. That's what we look for first. We can narrow our search to those places that are compatible with life as we know it on Earth. We can't discount, however, life based on totally different parameters, somewhere else.

There is a theory, proposed in 1962, that RNA worlds could exist, without DNA. The chemistry is different, and RNA as a precursor is the only process we have observed. RNA has, in the laboratory, been shown capable of synthesizing more RNA. RNA uses ribose instead of deoxyribose and its nucleo-bases substitute uracil for thymine. The current thought is that RNA preceded DNA. It is a somewhat more fragile molecule.

The big picture question is, "is life unique to Earth?" It only takes one data point to answer that in the negative. There's a whole universe of data points to examine. There is no compelling argument either for or against extra-terrestrial life. We don't know enough about our origins to suggest we are unique or not.

When we first search other areas for life, we assume it will be like that on Earth, so we look for places like Earth. We can expand that into the Habitable zone in our solar system, going out past Saturn. That's enough to do, so we're ignoring planets that are not in their own stars' habitable zone. We may have to go back and rethink that.

In the short term, we search for "life as we know it." We know the parameters. But we must also be ready to encounter alternative as we don't know it. Keep the options option. Darwin thought that life was formed by spontaneous generation, Pasteur said, Life from life. People have been thinking about these questions for a while

The Greek philosopher Epicurus around 300 BC said, "In all Worlds there are living creatures and plants."

Perhaps, the process to evolve lifeless matter into life is so complex, it only happened once. We'll never prove that there is no other life in the Universe. It only takes one case to prove there is. Meet the neighbors......

Author

The author has a BSEE in Electrical Engineering from Carnegie-Mellon University, and Masters Degrees in Applied Physics and Computer Science from the Johns Hopkins University. During a career as a NASA support contractor from 1971 to 2013, he worked at all of the NASA Centers. He served as a mentor for the NASA/GSFC Summer Robotics Engineering Boot Camp at GSFC for 2 years. He teaches Embedded Systems for the Johns Hopkins University, Engineering for Professionals Program, and has done several summer Cubesat Programs at the undergraduate and graduate level.

He began his career in Aerospace with Fairchild Industries on the ATS-6 (Applications Technology Satellite-6), program, a communication satellite that developed much of the technology for the TDRSS (Tracking and Data Relay Satellite System). At Fairchild, Mr. Stakem made the amazing discovery that computers were put onboard the spacecraft. He quickly made himself the expert on their support. He followed the ATS-6 Program through its operation phase, and worked on other projects at NASA's Goddard Space Flight Center including the Hubble Space Telescope, the International Ultraviolet Explorer (IUE), the Solar Maximum Mission (SMM), some of the Landsat missions, and others. He was posted to NASA's Jet Propulsion Laboratory for the MARS-Jupiter-Saturn (MJS-77), which later became the Voyager mission, which is still operating and returning data from outside the solar system at this writing.

He received NASA's Space Shuttle Program Managers Commendation award, two NASA Group Achievement Awards, and the NASA Apollo-Soyuz Test Program Award. He has completed over 42 NASA Certification Courses. He has led and supported efforts at all of the NASA Centers on terrestrial and planetary missions.

Mr. Stakem has been affiliated with the Whiting School of Engineering of the Johns Hopkins University since 2007. Mr. Stakem supported the Summer Engineering Bootcamp Projects at Goddard Space Flight Center for 2 years. He initiated a project that adapted NASA planetary image data processing applied to the Medical regime.

The Author is by no means fluent in biochemistry., as will be obvious from the rest of the text. He spent 42 years in all of the NASA centers, as an Engineer.

Mr. Stakem can be found on Facebook and LinkedIn. Comments, corrections, suggestions are appreciated.

Life

NASA's definition of life is based on what we know of life on Earth. We have, currently, no other examples. Here is the definition:

Recent study's have proposed the following necessary set of universal attributes of life:

1. life must exploit thermodynamic disequilibrium in the environment in order to perpetuate its own disequilibrium state
2. life most probably consists of interacting sets of covalently bonded molecules that include a diversity of hetero-atoms (e.g., N, O, P, S, etc. as in Earth-based life) that promote chemical reactivity;
3. life requires a liquid solvent that supports these molecular interactions; and
4. life employs a molecular system capable of Darwinian evolution.

These attributes imply the following basic universal functions:

1. life harvests energy from its environment and converts it to forms of chemical energy that directly sustain its other
2. functions, and thus, life requires usable sources of

energy;

3. life sustains "metabolism," namely a network of chemical reactions that synthesize all of the key chemical compounds that are required for maintenance, growth, and self-replication, and, thus, life needs chemical "building blocks" and an appropriate solvent to host these reactions; and

(3) life sustains an "automaton," a multi-component system that is essential for self-replication and self-perpetuation and, thus, life needs information-rich chemical compounds and favorable environmental conditions in order to sustain this complex machinery.

Earth Life

Life on Earth is incredibly diverse. Some of us breath air, some of us breath water. In both cases, we are seeking oxygen. All life we are familiar with is carbon based. It requires water. There are other bases possible. Pretty much everything we are familiar with is based on the dna structure, with 4 possible amino acid bases. The all of the life we know needs water. It uses RNA or DNA. It is formed around amino acids, which are organic compounds containing amine (NH_2) and carboxyl (C-O-O-H) building blocks.

Extremophiles are the outliers. They live in what we would consider extreme environments. It turns out, the smaller you are, the more likely you are to be tolerate of wide ranges of environment. Bacterial spores have been found that are 40 million years old (based on the rock they were found in.) Life has been found in the greatest depth of the oceans, and in a lake beneath Antarctica. Extremophiles scoff at these environments that would kill us quickly. They have also been fold inside rocks under the Atlantic seafloor. It seems their amino acids and their protein folding ability allow them to adapt to their conditions. Protein folding refers to a protein chain forming into a 3-dimensional structure. The position of the various amino acids making up the protein is the important part. Protein folding is a spontaneous reaction, which is dependent on temperature. The protein folding process has been extensively computer- modeled. The process that uses information from a gene in the synthesis of a function product is called gene expression. Messenger RNA conveys genetic information from DNA to a riibosome. This specifies the sequence of amino acids.

Let's look at the edges. Life has been found near geothermal vents under the ocean, perfectly happy at 120 degrees C. Another organism thrives in ice at -20C. Different variations of life at happy at pH values above 11 (akaline) and below 0.6 (acidic). Although complex molecules are disrupted by ionizing and UV radiation, certain organisms love it. Organisms in the Marianas Trench in the Pacific Ocean are happy with the 1,100 bar pressure.

Meteorites have been found with the building blocks of life embedded. The big question is, was this due to contamination? It has been shown that *left-handed* molecules (all Earth life has these) can be created in asteroids by a non-biological process.

There are disagreements on the definition of intelligence, but some of our fellow Earthlings exhibit what we recognize as intelligent behavior. This can take the form of swarm or hive intelligence.

Cetacean intelligence refers to the cognitive ability of whales, porpoises, and dolphins. These are all mammals, living in a water environment. Brain size used to be used as a measure of potential intelligence, but this has been abandoned as a metric. Otherwise, sperm whales would be more intelligent than humans. They might be. They are well adapted to their water environment, and can evidently communicate with each other. We can not yet communicate with dolphins or whales. Being able to do so would facilitate communication with other intelligent entities. Dolphins, on the other fin, have learned sign language, and can understand some whistles from humans.

Elephants are also considered to be intelligent, and are tool-makers. Other traits of intelligence are problem solving, ans self-awareness.. Dolphins, whales, and elephants are pod, or herd animals.

Swarms are collections of smaller co-operating individual; that can combine their efforts and work as ad-hoc teams on problems of interest. This behavior has been observed in birds, insects, and fish, as well as marine mammals. A collective behavior emerges from interactions between members of the swarm, and the environment. The resources of the swarm are organized dynamically.

Biological swarms, such as ants, achieve success by

division of labor throughout the swarm, collaboration, and sheer numbers. They have redundancy, as any individual can do any task assigned to the swarm. The individual units are highly autonomous, but are dependent on other members for their needs. They achieve success with a simple neural architecture and primitive communications.

Just as I was writing this, I saw an article that Russian cosmonauts found various Earth organisms were surviving on the outside of the International Space Station. Samples sent back to Earth indicated. these were not present of the outside of the station when is was launched. On a previous ISS resupply mission, bacteria accidentally went along for the ride, on tablet pc's. How these organisms got outside is not confirmed. Why they were able to survive for 3 years is not known. Current guidelines limit the biological contamination of space equipment to 500,000 spores, which represents the total in about 5 ml of seawater. There is a dedicated ESA facility outside the ISS for exposure of Earth organisms to the space environment, for eventual return to Earth for analysis.

The postulated set of conditions on Earth around 4 billion years ago is called the Primordial soup. This is part of the Heterotrophic theory of the origin of life, proposed by Alexander Oparin and John Haldane around the beginning of the 20th century. The process is in sync with Darwin, describing a process that goes from simple to complex. It assumes a primitive atmosphere with methane, water, and ammonia.

We make a first assumption that life in other places from Earth would us the same chemistry. That's a starting point, but who's to say the biochemistry has to be the same? On Earth, we are carbon-based lifeforms, using water as a solvent, and DNA and RNA. Alternate biochemistries might use silicon in place of carbon, ammonia, methane, or ethane in place of water. Of the various known molecules in the interstellar medium, 84 are carbon-based, and 8 are silicon based. Carl Sagan self-identified as a carbon chauvanist. Almost everything we know about life forms involves carbon, in our life zone of abundant water, and a fairly narrow temperature range, with liquid water. In other areas with different ecosystems, different chemistry's may be in use.

Other Life in our Solar System?

The Gaia hypothesis says that any planet harboring life will have an atmosphere in disequilibrium.

In our own solar system, there are planets and such that definitely do not harbor life, and those that could be a good candidate. We have visited all of the planets, at least with a fly-by, and quite a few of their moons. By direct sampling, we can search the candidates likely to harbor life. Included in this is the large asteroid Ceres, actually a dwarf planet in the asteroid belt. Ceres is know to have water ice beneath an outer crust. Water ice and a rocky surface are essential as a potential life habitat. Estimates for the habitable zone in our solar system goes from 0.38

to 10 AU. That includes everything out to Saturn.

Life can develop on a planet, or be transferred to it from another body. We don't have a clue which of these worked on Earth. We tend to anthropomorphize other planets, assuming their life will be similar to ours. That's a good starting point, for all we know right now. NASA considers currently that life requires liquid water, conditions that are favorable for the assembly and continuance of complex molecules, and appropriate energy sources. As we discover more and more exoplanets, outside of our home solar system, we have more places to look for life. Question number one is, do they have lift? Question number two follows on with, "is it the same as us?" These numbers change daily, but there are more than 4,000 known exoplanets, of which about half are in their star's habitable zone, as we define it for Earth life.

We make the assumption that life elsewhere uses carbon, hydrogen, oxygen, and nitrogen, as we have no counter examples. From these four elements, we can get amino acids, which can form proteins. Sulfur is required for proteins, and phosphorus is needed for DNA and RNA. The Miller-Urey experiment showed in 1935 that simple inorganic compounds in a primordial atmosphere can, with sufficient energy available, of amino acids. Good Starting point. What we don't want are energetic gamma rays, ionizing radiation, or excessive gravitational forces.

Mercury

Mercury, the closest planet to the Sun, is in tidal lock,

with one side always facing the Sun. Actually, there is a rare 3:2 spin-orbit resonance, not seen elsewhere. For every two revolutions around the Sun, Mercury rotates three times on its axis. It wobbles a bit, creating a twilight zone that is much less extreme. It has no known moons, or Trojans. Being so close to the Sun, it is difficult to observe the planet and its immediate vicinity. It has a heavily cratered surface. Mercury is a rocky planet, with long narrow ridges observed on the surface.

Mercury has a molten core with abundant iron. Its mantle is made of silicates. The crust is thought to be some 35 km thick.

The surface of the planet resembles Earth's moon, with large plains (mare) and craters. It is thought to be 4.6 billion years old. It has no atmosphere.

Mercury is currently being observed up close by the Messenger spacecraft, and this will increase our knowledge of the planet. We really don't have a way to image or study the sun-facing side. The spacecraft has seen evidence of more than 50 pyroclastic flows from active volcanoes.

The surface temperature of Mercury varies from -170 to 435 degrees Celsius. Water ice has been confirmed in deep craters at the poles, as well as on rocky planets around distant stars.

The eccentricity of Mercury's orbit is 0.2, the most of any planet in our solar system. The axial tilt is almost zero. Mercury's orbit is tilted 7 degrees to the plane of the ecliptic. The orbital eccentricity varies wildly from zero to 0.45 over a period of millions of years.

There is some indication that, since the planet wobbles a bit, there is a permanent, temperate "twilight" zone. The problem is getting there. There's no particular advantage to having humans live there.

Venus

Heavy greenhouse clouds of sulfuric acid trap solar energy, and cause massive global warming on a planetary scale. The surface temperature is high enough to melt some metals. We need to find out what went wrong on Venus, and try to avoid that on Earth.

Venus' atmosphere is 96% carbon dioxide at a surface pressure of nearly 100 times Earth's, a greenhouse gone wild. It has no moons. Venus is roughly Earth-sized, but something went terribly wrong

The heavily clouded atmosphere makes it difficult to observe Venus. We do know it rotates in the opposite direction to most of the other planets. It has no magnetic field.

Venus is a terrestrial, rocky planet, about the same size as Earth. It has a dense atmosphere of carbon dioxide. It's surface pressure is more than 90 times that of Earth's. Venus is hotter than Mercury in spite of being further from the Sun. It has seen extreme volcanism, but no lava flows have been observed. The surface was shaped by volcanic activity, and Venus has more volcanoes than Earth.

Venus orbits the Sun in about 225 (Earth) days, with a

small eccentricity. The planet has no moons. Due to extreme conditions, we can rule out life on Venus, as we know it, but we better check. There may be something lurking in the dense cloud cover.

Earth

Is there life on Earth. I believe so. But, could one of our exploration spacecraft detect it? The Galileo spacecraft used a loop around the Sun to gain velocity for its long trip to Jupiter in 1990. It looped back by the Earth again, and noted abundant gaseous oxygen, as well as atmospheric methane, and narrow-band pulsed radio transmissions. Ok, that worked. We just need to find them elsewhere.

Our nearest neighbor, the Moon, has almost no atmosphere, is irradiated constantly, and high large temperature extremes. It does seem to have sub-surface water ice. There is no observed pre-biotics in the lunar samples returned by the Apollo missions.

Near Earth Objects

Technically, an NEO is a solar system object whose closest approach to the Sun is 1.3 AU, and that comes in close proximity to the Earth There are 14,000 known asteroids in this category, 100 comets, solar orbiting spacecraft, and meteoroids. All these have the potential of striking the Earth. They are closely tracked from the ground, by NASA's Planetary Defense Coordination Office. A joint US/EU project called Spaceguard is tracking NEO's larger than 30 meters. Three NEO's have been visited by spacecraft.

Near-Earth Objects (NEO's), there more than 15,000. Stuff hanging around Earth. Chunks of rock. If these enter the atmosphere, they heat up and burn. Sometimes, enough is left to hit the ground, appearing as a rock. The easiest place to find meteors (as we call asteroids that hit the ground) is Antarctica, where they stand out against the snow and ice. A lot of the Antarctic meteors come from Mars, as one of my old professors proved. Martian Meteorite ALH84001 was found in the Allan Hills region of Antarctica. It was seen to have potential biosignitures of microbial life.

The Murchison meteorite was found in Australia in 1969. It weighed more than 100 kg. It was observed falling to the Earth, and fragments were found over an area of 13 square kilometers. It is a carbonaceous chondrite meteorite. Sixty-six amino acids were discovered in the meteorite. These particular amino acids have been created in the laboratory by an electrical discharge through a mixture of methane, nitrogen, ammonia, and water.

The Asteroid Belt, dwarf planets, and Centaurs

Asteroids have been imaged by the New Horizons spacecraft, on its way to Pluto, and by the Cassini spacecraft. The Pioneer-10 spacecraft was sent to study the far reaches of the solar system It passed through the Asteroid belt on its way to Jupiter and Saturn.

Although there are fewer than 10 planets, and less than

200 moons, there are millions of asteroids, mostly in the inner solar system. The main asteroid belt is between Mars and Jupiter. Each may be unique, and some may provide needed raw materials for Earth's use. There are three main classifications: carbon-rich, stony, and metallic.

The physical composition of asteroids is varied and poorly understood. Ceres appears to be composed of a rocky core covered by an icy mantle, whereas Vesta may have a nickel-iron core. Hygiea appears to have a uniformly primitive composition of carbonaceous chondrite. Many of the smaller asteroids are piles of rubble held together loosely by gravity. Some have moons themselves, or are co-orbiting binary asteroids. The bottom line is, asteroids are diverse.

Vesta and Ceres, large proto-planets in the asteroid belt, are also being searched for life, with the *Dawn* mission. Water vapor has been found., and definitive evidence of tholins on Ceres. Tholins are organic compounds can be synthesized from carbon dioxide, methane, ethane, nitrogen, and water. The impetus is solar ultraviolet radiation or energetic cosmic rays.

Exploring the known asteroids is a daunting challenge. On the other hand, the asteroids can be a significant source of raw materials for Earth. A conventional survey and exploration approach would take too long. What is needed instead is a multitude of autonomous and flexible nano-spacecraft. The architectural model is a swarm (social insect model) with distributed intelligence. Some asteroids have been seen to have traces of the amino acids adenine and guanine, and organic compounds. This raises the possibility that Earth was "seeded" from asteroids. It raises the question about the source. In addition, some asteroids are known to have water ice on the surface.

Compounds found on meteorites suggest that the chirality of life derives from abiogenic synthesis, since amino acid samples from meteorites show a left-handed bias, whereas sugars show a predominantly right-handed bias, the same as found in living organisms on Earth.

The asteroids are not uniformly distributed. In the asteroid belt, the Kirkwood gaps are relatively empty spots. This is caused by orbital resonance of the asteroids with Jupiter. Orbiting irregular shaped bodies is challenging, due to the irregular gravity field. This makes station keeping and attitude control a problem.

Centaurs are icy minor planets between Jupiter and Neptune There may be 44,000 others. Sometimes, they are captured by a planet's gravity, as a moon.

The Centaurs are icy minor planets between Jupiter and Neptune. A Centaur is a type of dwarf planet, orbiting the Sun. They are not quite making the cut to "real planet." There are 406 known. They have unstable orbits that intersect those of the gas giants. The largest known, Chariklo, has a ring system. These have not been photographed from a close position. Not much chance for life there.

Centaurs are in dynamic orbits, due to the gravity of their massive neighbour's. They act like a cross between a comet, and an asteroid. There is postulation that one of Saturn's moons may be a captured Centaur. A Centaur's orbit can be perturbed enough that it becomes a comet. Probably a write-off, hosting life.

There are currently 97 known Trans-Neptunian Objects that have not been otherwise classified. You never know.

The dwarf planets of our solar system include Ceres. Orcus, Pluto, Salacia, Varuna, Haumea, Quaoar, Makemake, 2007 OR10, Eris, and Sedna. Makemake has methane, ethane, acetylene, and tholinsThese smaller objects did not make the size cut to be a real planet. These all orbit the Sun. Orcus is a trans-Neptunian object, Salacia, Haumea, Quaoar, Makemake, and Varuna are Kuiper Belt objects. Eris is the largest of the dwarf planets, having its own moon. Sedna is beyond the Kuiper belt. It's orbital period "year: is 11,400 Earth years. It's in a highly elongated orbit, probably due to Neptune's gravity. Generally, a dwarf planet does not have enough gravity to clear its orbit of other material. Not all dwarf planets have yet been discovered or observed. There may be 10's of thousands. Judging by color, many of the trans-Neptunian objects have tholins.

Comets and Meteorites

There are some 5,253 known comets. The Deep Impact

mission returned images of the surface of comet Borrelly in 2001. That surface was hot (26-70C), dry, and dark. In July of 2005, the same mission sent a probe into Comet Tempel 1. It created a crater, allowing imaging of subsurface material. Water ice was seen. Comet Borrely has a coma, which proved to be vaporized subsurface water ice. Deep Impact went on to complete a flyby of Comet Hartley-2 in 2010. Cometary tholins were observed by NASA's Rosetta mission.

The 1999 Stardust mission retrieved sample material from the tail of Comet Wild 2 and returned it to Earth in 2006. It released a lander, Philae, which successfully touched down on the comet's surface in 2014.

Pioneer Venus observed Comet Halley while in transit. This was during a period when the comet was not visible from Earth, because of its proximity to the Sun. The Venus probe monitored the loss of water from the comet as it swung around close to the Sun.

Meteorites and comets have been to shown to contain over 100 organic molecules that are precursors to life. These include formaledhyde, ammonia, hydrogen cyanide and sulfide, methanol, acetylene, and carbon monixide, among others. Amino acids have also been found in meteroites.

Visitor from afar

Occasionally, our solar system get a visitor from afar, usually a comet, which makes a loop around the sun, and heads back out in space. Generally, our solar system bodies orbit the Sun in a disk called the ecliptic plane. Comets that are not necessarily in orbit around our Sun can take a path that are highly inclined to that plane. The first asteroid at a very high angle with respect to the ecliptic was recently observed. That means it did not originate in our solar system, but came from some where else in our galaxy, or beyond. There is no particular name for this class of objects, but the title "Exeroid" has been suggested. This will have to be cleared with the International Astronomical Union. The object was observed by a telescope on the Hawaiian mountain of Hakeakala. It is the first time an interstellar object has been observed. It was named Oumuamua, in Hawaiian, "messenger from a far, arriving first." It came from the direction of the constellation Lyra. Reviewing old data, 4 more similar objects have been spotted. They are not named yet, but are called 2011-SP25, 2017-RR@, 2017-SV13, 2018 EL6. These are currently between Jupiter and Neptune, in solar orbit, and will pass near Earth. They resemble centaurs, and are classified as counter-rotating, high inclination objects. Oumuamua seemed to have a parabolic orbit. It looped around and Sun and headed out of the solar system.

Mars

Mars, and its two tiny moons and seven Trojans has got some infrastructure in place – a communications relay and a weather satellite. There are several Rovers and landers on the surface.

The Viking program was a pair of spacecraft sent to Mars in 1975. Each spacecraft consisted of an orbiter, and a lander.

The Mars Pathfinder mission landed on Mars on July 4, 1997. It carried a Rover named Sojourner, which was a 6-

wheeled design, with a solar panel for power, but the batteries were not rechargeable. The rest of the lander served as a base station. Communication with the rover was lost in September.

The MER (Mars Exploration Rovers *Spirit & Opportunity*) are six-wheeled, 400 pound solar-powered robots, launched in 2003 as part of NASA's ongoing Mars Exploration Program. *Opportunity* (MER-B) landed successfully at Meridiani Planum on Mars on January 25, 2004, three weeks after its twin *Spirit* (MER-A) had landed on the other side of the planet. Both used parachutes, a retro-rocket, and a large airbag to land successfully, after transitioning the thin atmosphere of Mars.

The Spirit unit became stuck in 2009, and engineers were unable to free it after 9 months of trying. It was re-tasked as a stationary sensor platform. Contact was lost in 2010.

This is an ongoing mission. It was originally planned for 90 days, but the *Opportunity* Rover is still collecting useful data regarding potential life on our sister planet some 11 years later as of this writing. It has traveled over 35 kilometers on the Martian surface.

The Mars Science Laboratory landed successfully on the Martian surface on August 6, 2012. It had been launched on November 26, 2011. It's location on Mars is the Gale crater. It is designed to operate for two Martian years (sols). The mission is to determine if Mars could have supported life in the past, which is linked to the presence of liquid water.

The Rover vehicle *Curiosity* weights just about 1 ton (2,000 lbs, 900 kg.) and is 10 feet (3 meters) long. It has autonomous navigation, and is expected to cover about 20 km over the life of the mission. The platform uses six wheels

Communication with Earth uses a direct link, and via a relay spacecraft in Mars orbit. At landing, the one-way communications time to Earth was 13 minutes, 46 seconds. This varies considerably, with the relative positions of Earth and Mars in their orbits around the Sun.

The science payload includes a series of cameras, including one on a robotic arm, a laser-induced laser spectroscopy instrument, an X-ray spectrometer, and x-ray diffraction/fluorescence instrument, a mass spectrometer, a gas chromatograph, and a laser spectrometer. In addition, the Rover hosts a weather station, and radiation detectors. There is cooperation between in-space assets and ground rovers in sighting dust storms by the meteorological satellite in Mars orbit.

Curiosity's exploration of the ancient lake bed, known as Gale Crater resulted in some new discoveries that NASA released on June 7, 2018. It found organic molecules, particularly methane, below the surface. Curiosity has a sampling drill (that, unfortunately, is limited to 5 cm.), a mass spectrometer, and a gas chromograph. On Earth, most methane is from biological processes. It can be produced by in-organic processes, however. Scientists have also discovered a season pattern in the amount of methane in the atmosphere, in the amount of a factor of three, that may point to sub-surface storage.

The source of the Mars methane has to be resolved. This will be a major goal of NASA's and ESA's next landers. A large underground reservoir of methane could be very useful for return-trip rocket fuel. It would be more interesting if it was formed by a biological process.

NASA's Maven (Mars Atmosphere and Volatile EvolutioN Mission) mission to Mars is an orbiter, to study the Martian atmosphere It was launched in November 2013, and reached Mars in September of 2014. It is still operating as of this writing.

The ExoMars mission is a joint ESA-Russia project, specifically looking for life on Mars. The first part, the ExoMars Trace Gas Orbiter is already there and working. Unfortunately, its lander crashed. The second part, due to launch at the 2020 opportunity, is a landing platform with a rover. It will be actively searching for biosignatures.

We may have already contamiated Mars with Earth life clinging to the landers. Although they go through a strict decontamination procedure, following the Outer Space Treaty and the COSPAR guidelines for planetary protection. NASA uses the Management Manual NMI-4-

4-1, *NASA Unmanned Spacecraft Decontamination Policy*. This still allows for some Earth-based microbes. Until very recenting, it was thought that the environment at Low Earth Orbit (LEO) was entirely hostile to life. The radiation environment, the vast fluctuation of temperatures, the lack of nutrients would kill off anything. That all fell apart when Earth bacteria was seen to be thriving on the outside of the International Space Station. This is a game changer. Now we have to be even more careful in searching for lief on Mars, for example, that we don't seed the red planet. As the exploration of the outer planets's watery moons proceeds, we need a new approach to avoid contamination.

Some meteorites that hit Earth are from Mars. A lot of these land in Antarcticia. (One of my old college professors worked out the math.) In 1884, in the vivinity of the Allwn Hills in Antarcticia, a roughly 2 kilogram Mars meteorite was found. It was vey idfferent from previous Mars meteorites, and was siad to have evidence of microscopic fossils. This was disputed.

The Gas Giants and their moons.

The Gas giants are the planets Jupiter and Saturn. These are the responsibility of the Jet Propulsion Laboratory. These each have extensive ring and moon systems that have been imaged, but not yet explored.

Pioneer 10 was the first mission to Jupiter, followed by Pioneer-11 in 1973, and, as of this writing, there have been 8 total. Jupiter has a very high trapped radiation environment. They are mostly all different, and some are thought to be capable of hosting life, as we know it. The moon Io has volcanic activity, and Europa has water ice on the surface. Europa is considered "one of the most promising extraterrestrial habitable environments in our solar system" according to the most recent Planetary Society's Decadal Survey. A proposed mission, ExCSITE, would provide characterization of the surface properties.

NASA is looking at the Explorer CubeSat for Student Involvement in Travels to Europa (ExCSITE) Mission. This involves multiple Cubesat imagers and impactors.

The Voyager missions were originally termed the "Grand Tour" and were to have visited Mars, Jupiter, and Saturn, with possibly some of the outer planets as well. The mission was called MJS-77. Budget constraints caused the mission to refocus on Jupiter and Saturn alone. The author worked on this mission.

Jupiter

Jupiter has 79 known moons, and perhaps 1 million Trojans of 1 kilometer or larger. These tend to congregate at L4 and L5. The largest has a diameter of several hundred kilometers. The International Astronomical Union just announced as this book was being updated the discovery of 12 previously unknown moons of Jupiter, by an observatory high in the Andes in Chile. Only one has been named so far, Valetudo, a great-granddaughter of Jupiter The one way light time for Jupiter is 33-53 minutes.

Cassini observed the planet from close-up in the year 2000, and studied the atmosphere. Galileo entered Jupiter orbit in 1995, and returned data on the planet and the four Galilean moons until 2003. Three of the moons have thin atmospheres, and may have liquid water. The moon Ganymede has a magnetic field. Galileo was in the right place at the right time to see the comet Showmaker-Levy-9 enter the Jovian atmosphere, and launched an atmospheric probe.

The Hubble Space Telescope discovered that the satellite of Jupiter, Ganymede, has a large saltwater ocean, under an ice crust. This is now thought to be the best current hope for finding life on another planet.

The Juno mission to Jupiter has just arrived after 5 years of travel, and was getting settled in to begin its observations. This project was launched in August of 2011, and arrived at Jupiter in July 2016. It was placed in Jupiter elliptical polar orbit for 5 years, and will de-orbit into Jupiter in February 2018. This is to ensure burn-up of the spacecraft to avoid any biological contamination of Jupiter or its moons. It is scheduled to make 37 orbits. The orbit was chosen to minimize contact with Jupiter's intense trapped radiation belts. It's sensitive electronics are housed in "the Juno Radiation vault," with 1cm titanium walls. It will have available to it some 420 watts of power, from the solar arrays.

The Galilean moon Europa has water ice on the surface, and could be a prime candidate for a biosphere. Europa will be the target for a new NASA spacecraft in the 2020's, as part of the Oceans Worlds Exploration Program. The Europa Clipper mission will conduct follow up studies of the Ocean world around 2023. The Galileo spacecraft discovered the sub-surface ocean. In addition, NASA is looking at the Explorer CubeSat for Student Involvement in Travels to Europa (ExCSITE) Mission. This involves multiple Cubesat imagers and impactors.

ESA 's Jupiter Icy Moons Explorer, *JUICE*, is going to Jupiter, and will focus on the moons Europa, Ganymede and Callisto. These are all thought to have liquid water beneath their surfaces. This makes them potentially habitable environments. The mission is scheduled for launch in 2022, reaching the Jovian system in 2029. It will enter Ganymede orbit in 2032. At the same time, NASA's Europa Clipper mission will be in the Jovian system, orbiting the bog planet to make 45 close fly-bys of the moon. It is possible that Europa Clipper will carry several Cubesats which can be released into plumes from the moon's surface.

The four largest of Jupiter's moons can be seen from Earth with a modest telescope. They could qualify as dwarf planets, if they orbited the Sun, not Jupiter. The rest of the moons, and we have probably not found them all, are irregular, orbiting randomly. For life-hosting candidates, we have the Galilean moons, Io, Europa, Ganymede, and Callisto. The mission should reach Ganymede in 2033, and will orbit the moon.

Io is the 4th largest of the Galileo group, and has the highest density. It has the least water of any body in the solar system, seen so far. Over 400 volcanos have been observed. The crust of the moon seems to be silicate and iron. Due to gravitational forces from Jupiter and the other 3 Galilean moons, Io gets jerked around a lot, leading to tidal heating, and a lot of volcanism. Not our top candidate.

Europa is the smallest of the 4 Galilean moons, slightly smaller than Earth's moon. It has a water-ice crust, and most likely, an iron-rich core. It has a very smooth surface. It is possible that the moon has a sub-surface ocean, which is warmed by tidal flexing. Europa is a good candidate for consideration of life. and perhaps as a base for ring mining. Almost all of the moons are in tidal lock with Jupiter.

Ganymede has the distinction of being the largest moon in our solar system. It is a bit larger than the planet Mercury, but doesn't seem to have much of an atmosphere. It has a metallic core, and a magnetic field. Its composition is mostly silicate rock and water ice. It is estimated to have more water than Earth. There is a thin oxygen atmosphere. Pretty close to what we're looking for.

Callisto is also a possibility, and it is further from Jupiter and its trapped radiation belts. Colonies at these distances from Earth would certainly need to be self-sufficient, due to the long travel time involved.

Saturn

Saturn and it's 62 known moons has a one-way light time around 1.4 hours. Saturn has been visited by spacecraft four times. The first was a flyby by Pioneer-10 in 1979. This showed the temperature of the planet was 250 degrees K. Voyager-1 visited in 1980. It conducted a close flyby of the moon Titan to study its atmosphere. It is, unfortunately, opaque in visible light. We do know it rains methane. Voyager-2 swung by a year later, and data showed changes in the rings since its sister mission visited the year before. Temperature and pressure profiles of the atmosphere were gathered. Saturn's temperature was measured at 70 degrees above absolute zero at the top of the clouds, and -130 C near the surface. The flybys discovered additional moons, and small gaps in the rings. Organics have be observed at Titan, in January of 2019. It is known to have a liquid water ocean.

Cassini was the fourth spacecraft to study Saturn, which has rings, and is smaller than Jupiter. The rings were confirmed by the Voyager spacecraft in the 1980's. Cassini entered into Saturnian orbit, and is still returning data. The one-way communications time varies from 68-84 minutes. It has also collected data on the Saturnian moons Titan, Enceladus, Mimas, Tethys, Dione, Rhea, Iapetus, and Helene. Things are strange in the Saturnian system. Cassini observed a hurricane in 2006 on the planet's south pole. It appears to be stationary, 5,000 miles (8,300 km) across, 40 miles (67 km) high, with winds of 350 mph (560 kph). The large moon Titan has lakes of a liquid hydrocarbon, with possible seas of methane and ethane. Cassini launched a probe *Huygens* to Titan, and it landed on solid ground below the atmosphere. The Cassini mission was responsible for the discovery of seven new moons of Saturn. The moon Rhea appears to have tholins, as does Triton

The Voyager-2 Mission was the first to observe the moon's surface in detail. Cassini discovered a possible atmosphere on the moon Enceladus, with ionized water vapor, and ice geysers. It was once thought to be comprised entirely of water ice, but subsequent calculations of its mass and density preclude this.

Enceladus, the ocean moon of Saturn, has some of the building blocks of life – complex organic molecules, based on observations from the Cassini spacecraft. Only Earth and Enceladus satisfy the minimum requirements for life (as we know it). There is a global ocean, with observed plumes of water vapor. This is termed a cyro-volcano. Simple organic compounds with molecular masses less than 50 atomic mass units have been observed in the plume material. The subsurface ocean lies beneath an estimated 30 km of water ice. The depth of the ocean has been calculated to be around 30 km. It is speculated that a organic-rich film lies on the top of the water.

The Cassini spacecraft flew through flew through a gas cloud ejected from Enceladus with its mass spectrometer instrument. The cloud was mostly water vapor with nitrogen, methane, and carbon dioxide. It is conjectured that Enceladus is the source of Saturn's E ring. Further studies show the presence of additional hydrocarbons such as methane, propane, acetylene, and formaldehyde. It has been shown that the south polar region is heated from the interior of the moon. Additional fly-by's revealed the presence of ammonia. It is also geologically active, sending out plumes of ionized water.

Things are strange in the Saturnian system. Cassini observed a hurricane in 2006 on the planet's south pole. It appears to be stationary, 5,000 miles (8,300 km) across, 40 miles (67 km) high, with winds of 350 mph (560 kph). The large moon Titan has lakes of a liquid hydrocarbon, with possible seas of methane and ethane. The Cassini mission was responsible for the discovery of seven new moons of Saturn. It conducted a close flyby of the moon Phoebe, and two fly-bys of Titan. Phoebe has a retrograde orbit. Cassini launched the Huygens probe onto the moon Titan. It landed, and continued to supply data. Cassini was also responsible for discovery of another ring, as well as finding lakes of hydrocarbons near the north pole of Titan. These turned out to be larger than first thought, and we renamed seas. Keeping busy, Cassini discovered four additional moons in 2009. Lasting beyond its primary mission of completing 74 orbits, the spacecraft went into extended missions, operating until 2017, when it entered the Saturnian atmosphere.

As with Jupiter, Saturn itself is not a good candidate for hosting life, but some of its moons may be. The Cassini mission discovered large lakes or seas of hydrocarbons near the planet's north pole.

The Icy Giants *and their moons*

Uranus and Neptune differ from Jupiter and Saturn in that they have heavier volatile substances in their atmosphere, and are now referred to as ice giants. Their atmospheres are known to contain water, ammonia, and methane ice. Although the ice giants are outside of the habitable zone, we might find some surprises on their moons, especially the water-moons.

Uranus

Uranus has 27 known moons, a 13-ring system, and a one-way light time of 2.7 hours from Earth. Uranus was imaged in a flyby by the Voyager-2 spacecraft in 1986, which also captured some images of the Uranian moon Umbriel. Uranus and Neptune are two of the great remaining unknowns in the solar system, since neither have been explored in detail, by a dedicated mission. There is a desire to put an explorer spacecraft in orbit, and use that as a platform to launch probes into the atmosphere, and visit the moons.

Uranus has a strange predominately water-ammonia ocean, which is electrically conductive. A major targeted mission is the Uranus orbiter and Probe. Mission analysis comes up with a 12-13 year cruise from Earth to Uranus.

Uranus has thirteen inner moons, five major ones, and nine irregular moons. The inner moons have the same properties are the ring system. The major moons show signs of volcanism. The irregular ones have strongly inclined orbits, some retrograde. The major moons are roughly in the plane of the planet's equator.

Five moons would be considered planets if they orbited the Sun, not Uranus.

The moon Titania is Uranus' largest, and probably has a rocky core, with surface ice. It is about one half the size of Earth's moon. There may be liquid water under the ice. The surface shows impact craters, and there is frozen carbon dioxide on the surface.

The moon Oberon is the outermost major moon also probably has a rocky core and an icy covering, with liquid water at the boundary. This would require ammonia or an other anti-freeze to be present. There is evidence of asteroid impacts on the surface.

Arielseems to be made up of water ice and a dense, non-ice element such as rock. The surface ice is crystalline, and there is evidence of carbon dioxide. Some of the surface is cratered.
Umbriel

The composition of Umbriel is similar to that of Ariel. It also has visible impact craters.

Miranda is the smallest of the five round satellites, with a diameter of 475 km. It has a very low density, that may be due to a 60% composition of liquid water. It has a rocky surface. Water has been detected on the surface.

A future planned mission is the Uranus Orbiter and Probe. There is a planned launch in 2030, with Uranus orbit insertion in 2041. The mission is named Oceanus. (Origins and Composition of the Exoplanet Analog Uranus System).

Neptune

Neptune has 14 known moons, and 18 known Trojans. It's one-way light time is around 4.3 hours. Neptune has also been visited by Voyager-2 in 1989. It discovered six new moons. That is the extent of close-up observations of the planet. Neptune has rings, like Jupiter and Saturn, and a great dark spot. It's moon Triton has geysers and polar caps. Triton has an interesting retrograde orbit – it goes in a different direction than the other moons. Triton's surface is mostly frozen nitrogen, and is geologically active. It is speculated that Triton has a subterranean ocean. The moon Ptoteus is an ellipsoid, not a sphere.

The moon Triton has an interesting retrograde orbit – it goes in a different direction than the other moons. Triton's surface is mostly frozen nitrogen, and is geologically active. It is speculated that Triton has a subterranean ocean. The moon Ptoteus is an ellipsoid, not a sphere. The atmosphere is mostly hydrogen and helium, with some hydrocarbons. It also has water, ammonia, and methane ice.   Neptune also has a significant magnetosphere, that interacts with the solar wind.

Now that Pluto has been downgraded to minor planet status, Neptune is the farthest planet in our solar system from the Sun. It is at an average distance from the Sun of 30 AU, and orbits every 165 years.

Pluto, and beyond

Pluto was downgraded from a planet to a Kuiper Belt object. The Kuiper belt is a circumstellar disk, extending from Neptune (30 AU) to 500 AU. It resembles the asteroid belt. It holds mostly small bodies. Pluto's moon Charon has evidence of tholins, as does one of its moons, Charon.

There is lots of interesting stuff beyond Pluto, before Inter-stellar space is reached. The minor planet 90377 Sedna is located at about 86 AU and is one of the Trans-Neptunium Objects. Pluto is far beyond the habitable zone, and unless there is a Jupiter-sized planet with moons we have missed, there not much chance in finding life that far out from the sun.

The dwarf planet Eris has the distinction of being the most massive known dwarf planet. It does orbit the sun, and is classified as a Trans-Neptunian Object. It is slightly more massive than Pluto.

NASA's Ocean Worlds Exploration Program

At the moment, the best place to search for other life in our solar system seems to be the Ocean Moons. This includes Titan, Enceladus, and Europa. There is also a defined NASA Mission called Europa Clipper, which will study the Ocean moon from Jupiter Orbit. Following that might be the Europa Lander, and a mission to Neptune's moon Triton.

The Ocean Worlds Exploration program will look at solar system bodies that are currently known to have subsurface oceans. This includes Europa, Enceladus, and Titan. The Europa Clipper is the first explorer of this mission.

Other solar system objects that are suspected of having water are Ganymede, Dione, Pluto, Triton, Ceres, Callisto, and Mimas. Earth itself is an Ocean world.

One precursor for life (as we know it) is liquid water. This dictates certain temperature ranges, as well as an atmosphere to keep the water in place. That, in turn, dictates a certain planetary mass, with enough gravity to hold the atmosphere. Another nice thing to have is a planetary magnetic field, to deflect charged particles from the Sun. A few energetic particles may be useful to "stir the mix."

One excellent idea to demonstrate whether we can detect life at a distance was suggested by Astrophysicist Carl Sagan during the Galileo mission to the outer planets. He suggested we look back at Earth, and see if we could detect life, where we knew it existed.

The Search for Extra-terrestrial Life beyond our Solar system.

This is a lot harder, due to the distances involved. We know of the existence of Exo-planets, mostly by their effects on their "Suns." We know from Earth the temperature and pressure limits for life. If we are looking for Earth-like life, we can apply those parameters for our searches. This allows us to characterize the Ex0-planets, and examine those most like Earth, in the Goldilocks zone of their stars. On the other hand, if ex0-life is nothing like ours, the we are missing something. Besides Exo-planets, we know of exo-solar systems consisting of multiple planets around other stars, and we know of exo-moons orbiting exo-planets.

Nowhere is it written that life only exists on Earth. Figuring out if that is true is the challenge.

In the year 2,000 the discovery of exoplanets began for real, as new technology was deployed. Older programs, such as SETI (Search for Extra-Terrestrial Intelligence) had been searching the radio spectrum for some time. There are more un-inhabited planets that we know of (3,874) than those with radio transmitters. exoplanets come in all sizes, and about 1 in 5 stars like our Sun have one similar to Earth. Do the math. With 200 billion stars in the Milky Way along, that could mean potentially 11 billion potential Earth's out there. On the average, each star has one planet. In addition, about 1 in 5 stars like our Sun, have a planet in the "curiosity zone." In a binary star system, an exo-planet can orbit either or both stars. There are exoplanets in triple star systems, and at least one has been observed to have a ring system like Jupiter and Saturn. Exo-moons orbit exoplanets. Atmospheres have been discovered around some exoplanets. Some exoplanets are tidally locked to their primary, much like Mercury in our system. Exoplanets can be "rocky" like Venus, Earth, and Mars, or "gassy," like Jupiter and Saturn.

A few thousand have been cataloged. Can exoplanets harbor life? To have the potential of life, the planets have to orbit the right type of star, at the right distance, called the Goldilocks zone (from the Fairy tale). Not too hot, not too cold, just right.

What exactly can we look for on planets of other stars? Biosignatures and Technosignatures. A biosignature is an indication of life chemicals, such as DNA. We have to be carefule in assuming that life elsewhere, if it exists, uses the same inputs and processes as that on Earth. For example, we know there is methane on Mars, but we don't know if it was produced by a biological process.

A technosignature would be an indication of intelligent life. This might include radio or light beacons or transmissions, deliberated sent, or "leakage" from a technology-savvy planet. We might also look for "alien" spacecraft, unusual travelling objects. A recent one of these, Oumuamua, (discussed previously), got scientists excited. NASA has held Technosignatures Workshops.

NASA's approach is the Astrobiology Strategy document. The current version is from 2014, with the previous being in 2008. This document sets the specific goals in astrobiology, from the overall objectives in planetary science. One key question is, what makes a planet habitable, or friendly to life. The three big questions discussed are:

1. How does life begin and evolve?

2. Is there life elsewhere?

3. What is the future of life on Earth?

Since we only have one example of a planet harboring life, the first approach is to find it elsewhere, if possible.

Life is so diverse on Earth, consider plants versus animals, had some none-of-the above, we may not recognize life when we se it.

The primary search regions are Mars, and the icy moons of Jupiter and Saturn.

NASA has defined six major topics of research:

1. Identify abiotic sources of organic compounds. Where did life's basic building blocks come from, and are they unique? Are there different biologies in different environments, using unique processes? We have to keep in mind that the environmental conditions on Earth has changed drastically over the life of the planet, and conditions today are very different. What we seen in amino acids and dna might be related to environments that no longer exist. Also, elements necessary for the emergence of life may no longer be relevant, but still present. To deal with a changing environment, life has evolved to change was well.

2. Define   the   synthesis   and   function   of

macromolecules in the origin of life. Macromolecules include proteins and nucleic acids. Are these unique? Our "alphabet" of 20 amino acids leads to the protein architectures of Earth life. There are more amino acids, but they don't take a part in Earth-based biology.

3. Trace early life through increasing complexity in the four billion years it has been around. There is a lot of diversity in Earth life. Also, there is some question about what constitutes life, such borderline cases as viruses and prions.

4. Define the co-evolution of life and its physical environment. Scientists are trying to define the properties of Earth life that are essential, and explore any possible alternatives, and the environments they require.

5. Identify, explore, and characterize environments for habitability and bio-signatures. As an example, Mars has methane, a gas that can be produced by life, or simple chemistry.

How are habitable worlds constructed? Habitability has other issues. For example, in our solar system, the Earth is the only "habitable" planet for us. Some of the icy moons of Jupiter and Saturn might be habitable for other forms of life.

A new tool in the search for Exo-planets was launched in 2022. The James Webb Space Telescope was put into its orbit, after some 25 years of design and testing. As of this writing, it is still running on-orbit testing and calibration.

It sees in the infrared spectrum, where HST is a visible light telescope.

The Hubble Space Telescope was launched in 1990. It is used actively for discovering exo-planets.

Panspermia

Panspermia is a hypothesis that life is distributed throughout the universe, distributed by space dust. This supposed that the building blocks of life can survive direct exposure to interstellar and intergalactic space. The supposition is in opposition that life can evolve anywhere in the universe, although both can be true. Since we haven't found extra-terrestrial life and since we don't know the origins of life on Earth, we can't say. The Panspermia theory was discussed as early as 5th century BC in Greece. The theory of Life from non-life can be traced back to Aristotle in Greece. The idea of spontaneous generation of life, popular in the middle ages, was kicked out by Louis Pasteur

Complex organic molecules have been observed (by their spectra) in interstellar space. This leads to a theory that life from one place in the Universe can migrate to another. Interstellar space is not as inhospitable to life as once thought. This allows for either life developing in multiple locations simultaneously, or developing in one location and traveling. Both cases can be supported. But, keep in mind, we are extrapolating from one known datapoint. Life on Earth is considered to have started as bacterium. We have evidence of them from greater than 3.5 billion (10^9) years ago. These are micro-fossils from

Northern Quebec, Canada.

Abiogenesis

Abiogenesis is the process by which life has developed from non-living matter. We do not understand all the details of this process, to put it simply. We have seen large building blocks of life self-organize from chemicals in their environment. Life implies complexity. We need, at the molecular level, self-replication and self-assembly. It is a continuing process of increasing complexity. What is not known is where is the magic moment when chemicals organize themselves into something we could call "Life"?

Most life as we know it is carbon-based. Using the chemistry of carbon and water/ there are four key groups of chemicals: lipids, providing fatty cell walls; carbohydrates such as sugar and cellulose; amino acids, for protein metabolism; and nucleic acids, RNA and DNA. One of the key questions is how self-replicating molecules came about. This is also a debate about DNA life, like ours, and potential RNA-based life.

In 1952, the Miller-Urey experiment demonstrated that amino acids can be synthesized from inorganic compounds. Regardless of origin, Life has features of self-organization and self-replication.

The Drake Equation

The Drake equation, by Dr. Frank Drake in 1961, is a way of estimating the number of extraterrestrial

Civilizations that can communicate. The equation multiplies these factors:

R,∗ the average rate of star formation in our galaxy.
fp, the fraction of those stars that have planets.
ne, the average number of planets that can potentially support life per star that has planets.
fl, the fraction of planets that could support life that actually   develop life at some point.
fi, the fraction of planets with life that actually go on to develop   intelligent life.
Fc, the fraction of civilizations that develop a technology that   releases detectable signs of their existence into space.
L, the length of time for which such civilizations release detectable signals into space.

In 1961, guessing at the parameters, the value for N was 10^3-10^8. As more data is collected, better values for the parameters evolve. The current value is around 15 million.

The resultant figure for the universe suggests it is highly unlikely that Earth hosts the only intelligent life that has ever occurred. Maybe we can check with our neighbors...

The Fermi paradox, named after physicist Enrico Fermi, is the apparent contradiction between the lack of evidence and high probability estimates for the existence of extraterrestrial civilization. What are we missing?

Initially, we had to guess at most of these, but better observation has brought about better results. As our observation technology gets better, we get better numbers for the factors in the equation.

In 1992, the first discovery that our solar system was not unique in the Galaxy was made. This gave us some real numbers to plug into the Drake equation. This observation was done by radio-telescope.

The Red Dwarf star Trappist has a system of 7 planets, nearly all Earth size. That a red dwarf would have a planetary system was a surprise, but opened up the options for further research. At the moment, we can only detect exo-planets of Earth-size or larger.

SETI

Around 1900, astronomers began to search for radio signals from other than Earth. The famous inventor Tesla in 1896 thought his wireless equipment count contact intelligent life on other planets. This was thought feasible by radio pioneer Marconi. It went so far that, in 1924, a National Radio Silence Day was defined for a 36-hour period, with no radio broadcasts for 5 minutes of every hour. At the Naval Observatory in Washington, D. C. a three kilometer antenna was hoisted by a dirigible. The Chief of Naval Operations had a navy cryptographer on-site to translate messages from the Martian.

Early SETI work was done by Frank Drake (of the Drake equation) in the 1960's from the Greenbank radios Astronomy facility in West Virginia. This was called Project Ozma. NASA funding came around 1970, and the Deep Space Network routinely listens for signals today. A lot of the data from the radio-telescopes in put up on the internet, allowing people to sift through it, and participate in the program. This is called SETI-at-home.

SETA is a xenoarcheology project to search for alien artifacts. The term was invented in the 1980's.   It is based out of the Space Science Lab of the University of Berkeley in California. The program began in 1999, applying distributed computing to the task of finding relevant signals in mountains of data from radio telescopes at Greenbank and Arricebo. The program has downloadable search software called SETI-at-home to run on a user's computer. To date, no significant signals have been encountered. But, keep looking. Arecibo, for examples, generates about a terabyte of data per day. There are more than 5 million participants.

As a side point, Arecibo, in 1974, transmitted a message out towards the Globular Cluster M13. It consisted of 210 bytes. The digits from an image, partially developed by Dr. Frank Drake.

The sent message has 7 parts:

1. The numbers one (1) to ten (10)

2. The atomic numbers of the elements, hydrogen, carbon, nitrogen,oxygen, and phosphorus, which make up deoxyribonucleic acid (DNA).

3. The formulas for the sugars and bases in the nucleotides of DNA.

4. The number of nucleotides in DNA, and a graphic

of the double helix structure of DNA.

5. A graphic figure of a human, the dimension (physical height) of an average man, and the human population of Earth.

6. A graphic of the Solar System indicating which of the planets the message is coming from.

7. A graphic of the Arecibo radio telescope and the dimension (the physical diameter) of the transmitting antenna dish.

(courtesy, Wikipedia)

Search in our Solar System

In our own solar system, there are planets and such that definitely do not harbor life, and those that could be a good candidate. We have visited all of the planets, at least with a fly-by, and quite a few of their moons. By direct sampling, we can search the candidates likely to harbor life. Included in this is the large asteroid Ceres, actually a dwarf planet in the asteroid belt. For exo-planets, we can observe them with ground-based and space-borne telescopes, but we have little hope of visiting them in the near future. In interstellar space, the Voyager Spacecraft are still operating, as well as New Horizons. In a few million years, we may have an answer.

One precursor for life is liquid surface water. This dictates certain temperature ranges, as well as an atmosphere to keep the water in place. That, in turn, dictates a certain planetary mass, with enough gravity to hold the atmosphere in place. Another nice thing to have is a planetary magnetic field, to deflect charged particles from the Sun.

Mercury is too close to the Sun, and is in tidal lock, which means one side always faces the Sun, the other, deep space. However, the planet wiggles a bit, creating a Goldilocks zone which could possible harbor life.

Venus is an ecological disaster. The heavy cloud cover of greenhouse gases traps heat, making the surface hot enough to melt some metals. It might be possible for organisms, or life building blocks to exist high in the Venetian atmosphere.

We have been looking for life on Mars since the Viking program in 1975.

Curiosity's exploration of the ancient lake bed, known as Gale Crater resulted in some new discoveries that NASA released on June 7, 2018. It found organic molecules, particularly methane, below the surface. Curiosity has a sampling drill (that, unfortunately, is limited to 5 cm.), a mass spectrometer, and a gas chromograph. On Earth, most methane is from biological processes. It can be produced by in-organic processes, however. Scientists have also discovered a season pattern in the amount of methane in the atmosphere, in the amount of a factor of three, that may point to sub-surface storage.

Private Search

Breakthrough Initiatives is a private program, searching

for extraterrestrial life. Stephen Hawking helped organize the program in 2015.. The approach takes several forms, searching for radio or laser signals, sending messages, and sending a swarm or probes to the nearest star (Alpha Centauri) at 20% of the speed of light. These would be propelled by light sails, and would take an estimated 20 years to reach our neighbor. It would take a radio or laser message 4 years to get back to Earth. There would be a thousand tiny spacecraft, each weighing several grams. These would be taken to orbit on a mothership, and deployed,. A ground-based array of lasers would be used to accelerate the spacecraft, reflecting off of their 5 meter diameter sails.

Breakthrough Listen is a private initiative to search for intelligent life via communications. The project kicked off in 2016, and is based at the Berkeley SETI Research Center. It is focused on communications in the radio and optical frequencies. Data are collected at 24 gigabytes per second, and are analyzed by a compute-cluster. The data are available in an open data archive. The project is funded at the $100 million level. It is spearheaded by Frank Drake, chairman emeritus of the SETI Institute. Most of the data comes from the Greenbank Radio Telescope facility in West Virginia.

ExoPlanets

ExoPlanets are planets of other stars. Although it is difficult to see them through a telescope, we can define them by their gravitational effects on their primary (star).

A Galaxy, like our own Milky Way, is a collection of

stars gas, dust, and dark matter, gravitationally bound together. The objects orbit the galaxy's center of mass, and can be of an elliptical, spiral, or irregular shape. Between the stars and other objects in a galaxy is a gas, with a density around 1 atom per cubic meter. There are estimated to be around 10^{11} - 10^{12} galaxy's in our observable universe, each having 10^8 to 10^{14} stars.

At the moment, we know of more than 4,000 planets orbiting other stars. We don't know how many are in the habitable zone. Even the definition of "habitable" is not well defined, and goes according to "life as we know it." Exoplanets can be in any orbit – they can rotate in the same direction they are moving in orbit, or the opposite way (retrograde). There are also rogue planets, that do not orbit a star, but rather the galactic center. It is thought that there may be a billion of these in the Milky Way galaxy alone.

More than one exo-planet can orbit a distant star, giving us exo-solar systems. In addition, an exo-moon has been discovered orbiting a known exoplanet.

Search for Exoplanets

It is difficult to search for Exoplanets from the Earth's surface. It can be done with radio-telescopes. We need to search in the Infrared bands, which are mostly blocked by the atmosphere. Infrared can show us features of distant stars, but most of the visible light is scattered by space dust. Another good idea is simultaneous observations by different observatories with different point of view. Multiple spacecraft missions are dedicated to searching for planets around other stars, particularly those in the habitable zone. Some examples will be discussed. Wikipedia maintains an up-to-date list of potentially habitable exoplanets.

The Drake Equation

The Drake equation, formulated by Dr. Frank Drake in 1961, is a way of estimating the number of extraterrestrial Civilizations that can communicate. It was formulated by Dr. Drake at the Greenbank Radio observatory in West Virginia. The equation multiplies these factors:

R*, the average rate of star formation in our galaxy.

fp, the fraction of those stars that have planets.

ne, the average number of planets that can potentially support life per star that has planets.

fl, the fraction of planets that could support life that actually develop life at some point.

fi, the fraction of planets with life that actually go on to develop intelligent life.

fc, the fraction of civilizations that develop a technology that releases detectable signs of their existence into space.

L, the length of time for which such civilizations release detectable signals into space.

N , the number of civilizations in our Galaxy that ca communicate.

In 1961, guessing at the parameters, the value for N was

10^3-10^8. As more data is collected, better values for the parameters evolve. The current value is around 15 million.

The resultant figure for the universe suggests it is highly unlikely that Earth hosts the only intelligent life that has ever occurred. Maybe we can check with our neighbors...

The Fermi paradox, named after physicist Enrico Fermi, is the apparent contradiction between the lack of evidence and high probability estimates for the existence of extraterrestrial civilization. What are we missing?

Initially, we had to guess at most of these parameters, but better observation has brought about better results. As our observation technology gets better, we get better numbers for the factors in the equation.

In 1992, the first discovery that our solar system was not unique in the Galaxy was made. This gave us some real numbers to plug into the Drake equation. This observation was done by radio-telescope.

The Red Dwarf star Trappist has a system of 7 planets, nearly all Earth size. That a red dwarf would have a planetary system was a surprise, but opened up the options for further research. At the moment, we can only detect exoplanets of Earth-size or larger.

Dr. Rene Doyon of the University of Montreal, with an International Team, photoed three XO planets orbiting the same star. This is the first detected X0 solar system. He used the Keck and Gemini North telescopes in Hawaii, on the summit of the volcano Mauna Kea. Both of these have 10 meter primary mirrors.

NEXSS

The Nexus for Exoplanet System Science is a virtual institute focusing on the search for life on other planets. There are 16 science teams from ten universities, NASA Ames, the NASA Exoplanet Science Institute, and the Goddard Institute for Space Study's. The observed data from the Kepler planet-hunter spacecraft is available to the public. At this writing, here are more than 4,050 confirmed exo-planets.

Habitable Exoplanet Imaging Mission

The HabEX is a concept for a space telescope that is designed to image Earth-sized planets in the habitable zone of their Sun-like star. It is proposed, and may be chosen to implement, in 2020.

HabEx would be specifically looking for biosignature gases in the exoplanet's atmosphere. Some of these are oxygen, ozone, and water. Included would be carbon dioxide, carbon monoxide, and tetraoxygen O_4). Methane (CH_4) is also associated with life. The oxygen would need to be shown to originate in abiotic processes.

Ross-128b

Ross-128b is a confirmed exoplanet in the habitable zone of the star Ross-128. That star is about 11 light-years away. The planet was discovered in decades of data from the European Southern Observatory in Chile. It is thought that the planet is a rocky world. It does not transit its "sun," so no-traditional methods were used to surmise its existence. Also, no current telescope can discern its existence. Astronomers were able to discern the presence and abundances of several chemicals, from spectroscopy. These included iron, carbon, magnesium, aluminum, calcium, potassium, and titanium. It's host star is smaller than our sun, but the planet orbits closer, so it falls in a temperate zone, possibly conducive for life. It is one of the most Earth-like exoplanets found to date.

Alpha Centauri's planets

A special target for looking for exoplanets is Alpha Centauri, the Stat closest to out solar system, a mere 4.37 light years distant. It is actually a binary star system of Alpha Centauri A and Alpha Centauri B. There is an associated red dwarf, about 13,000 AU distant, called Alpha Centauri C, or Proxima Centauri. The European Southern Observatory discovered an Earth-sized planet orbiting Proxima Centauri in the habitable zone. This discovery came in 2016. Another small planet, smaller than Mercury, had been found in 2013, around Alpha Centauri. In 2015, the HST saw a transit event caused by a planet of about Earth's size. There may be more.

Kepler-443b

To date, the most distant confirmed found is Kepler-443b, 2,540 light years from out solar system. It seems to be in its primary's habitable zone, but has only a small chance of being rocky.

WASP-121b

The exo-planet WASP-121b orbits the star, WASP-121, some 850 light-years from Earth. The exo-planet is the first to have water in its stratosphere. It is a large planet, some 1.8 timers Jupiter's size.

Afterword

No matter how Life got going, we are frustrated by not knowing if we have neighbors. If Life is a one-time event, on Earth only we will be disappointed. But, it's a big Universe out there. We only know a portion of our own neighborhood. It would seem unlikely that we are that unique.

Glossary of Terms

Abiogenesis – process by which life originates from non-living matter.

Achiral – object that cannot be distinguished from its mirror image.

Acetylene – hydrogen carbon gas, C2H2.

Amination – amine groupt introduced into an organic molecule.

Amino acid – organic compounds – building blocks.

amu – atomic mass unit

ASIN – Amazon Standard Inventory Number.

Astrobiology – a branch of biology concerning life in the Universe.

Astrometry – measurements of positions and movements of objects in space.

Asteroseismology – study of oscillations in stars.

ATA – Allen Telescope Array, California.

ATP – Adenosine Triphosphate, energycarrier molecule in life.

AU – astronomical unit of distance, average distance from the Earth to the Sun, about 93,000,000 miles.

Autocatalysts – self-replicators.

Bacteria – small living biological cell.

Bar – metric unit of pressure, about 14.5 psi.

Biopoiesis – living matter evolving from relicating but non-living molecules.

Biosignature – evidence of past or present life.

Bracewell Probe – autonomous interstellar probe to communicate with aliens. Concept.

CAESAR – Comet Astrobiology Exploration Sample

Return.

Carbonaceous chondrites – class of non-metallic meteorites.

Cetacean intelligence – cognitive ability of whales, purposes, and dolphins.

CHAR - counter-rotating, high inclination (object).

Chemo-fossil – remains f a complex organic molecule.

Chirality – distinguishable from its mirror image.

CHNOPS -carbon, hydrogen, nitrogen, oxygen, phosphorus, sulfur,   combinations make up most biological molecules on Earth..

Chondrite – stony, non-metallic meteorites.

CHZ – circumstellar habitable zone.

Codon: a sequence of three DNA or RNA nucleotides that   corresponds with a specific amino acid. Acts to stop protein synthesis.

COSPAR – Committee on Space Research (International).

CRHI - counter-rotating, high inclination (objects).

Cyano – color blue; cyanide.

Cyanobacteria – blue bacteria that use photosynthesis and produce oxygen.

Dipeptide – organic compound formed from two amino acids.

DNA - Deoxyribonucleic acid, the double helix of genetic instructions.

DNA replication – process by which DNA produces two identical copies of itself.

EJSM-Laplace - Europa Jupiter System Mission Laplace; ESA, canceled.

ELF - Enceladus Life Finder.

Elsah – Encaladus life signatures & habitability

Enceladus – ocean moon of Saturn.

EnEx - Enceladus Explorer.

Enzymes: metabolic catalysts.

ESA – European Space Agency.

ESI – Earth Similarity index.

Ethylene – Acetylene C_2H_2.

Eukaryote: an organism whose cells contain a nucleus and other structures that are enclosed within membranes.

Exeroid – body from outside our solar system.

Extra-terrestrial – not on Earth

Exo-moon – Moon not around a planet in our solar system.

Exoplanet – Planet orbiting a Sun other than our own.

Exo-Solar System – Solar System beyond ours.

Extremophile – organism thriving in extreme environments.

Genetic code – set of instructions used by living cells to translate information.

Goldilocks Zone – area that is neither too hot nor too cold; from a Fairy Tale.

HabEx - Habitable Exoplanet Imaging Mission.

Habitable zone – range of orbital distances around a star where a planet can have liquid water on the surface.

HEIM - Habitable Exoplanet Imaging Mission.

HST – Hubble Space Telescope

Ice giant – large planet consisting of ices of various substances – Uranus and Neptune.

ICSU – International Council for Science.

IRM – interstellar radio message.

ISBN – International Standard Book Number.

Isomerization - the process by which one molecule is transformed into another molecule through the rearrangement of atoms.

JET - Journey to Enceladus and Titan.

JIMO – Jupiter Icy Moons Explorer.

JUICE – Jupiter Icy Moons Explorer.

Kardashev scale – measurement of a civilization's level of technology achievement.

LGM – little green men.

Lipid – biomolecule that is soluble in a non-polar solvent.

LUCA – last Universal common ancestor.

Macromolecule – large, complex polymer.

Metabolism – life-sustaining chemical reactions.

METI – messaging extra Terrestrials

Methanogens – microorganisms that produce methane.

MOMA – Mars Organic Molecule Analyzer, part of the ExoMars rover.

Moon – smaller astronomical body in orbit of a planet.

Moonmoon – a body revolving around a moon.

mRNA – messenger RNA.

MSR – Mars sample return mission.

NEXSS - Nexus for Exoplanet System Science (NASA).

NOM – natural organic matter.

Nucleobases – cytosine [C], guanin [G], adenine[A] or thymine [T]

Organile – special unit within a cell with a specific function.

Organic – pertaining to a living organism, currently,

carbon-based.

OWEP – (NASA) Ocean Worlds Exploration Program.

Panspermia – theory that life exists throughout the Universe.

Peptide – short amino acid chains linked by peptide bonds.

PH – percent hydrogen, measure of acidity. Log scale.

Planet – a body orbiting a star.

Polymer – large molecule.

Polypeptide - long, continuous, and unbranched peptide chain.

Prebiotic – nature environment chemistry before the advent of life.

Premordial soup – hypothetical conditions on Earth 4 billion (10^9) years ago.

Prokaryotes – unicellular organism, without a membrane-bound nucleus.

Protein – large long chain bio-molecules, formed from amino acids.

Protein folding – process for a protein chain to achieve a 3-dimensional structure.

Prion – a misfolded protean. Defective.

PSI – pounds per square inch; unit of pressure.

Ribosone – responsible for biological protein synthesis.

RNA - Ribonucleic acid, essential in biological processes.

SETA - Search for Extraterrestrial Artifacts.

SETI – Search for extra-terrestrial intelligence.

SETI-at-home – Internet-based collaborative seti.

Solar System – A star and its associated planets and such.

Spindle cells – neurons without extensive branching.

Found in humans. May be a marker for "intelligence."

TALISE - Titan Lake In-situ Sampling Propelled Explorer.

TBD – to be determined.

Telenomic – in biology, the principle that the body's structures and functions serve an overall purpose.

Terraforming – changing a planet or moon to be more Earth-like.

TESS – Transiting Exoplanet Survey Satellite.

Techo-signature -

Tholin – organic compounds formed by UV radiation from simple carbon compounds.

Tidal lock – where the same side of a object always faces the primary it is orbiting.

TIME – Titan Mare Explorer.

Trna – transfer RNAS

UV – ultraviolet.

Waterhole – frequency between that of H and OH.

WoW signal – strong one-time radio signal in 1977.

Xenoarchaeology – archaeological study of alien civilizations. No examples yet.

References

Anderson, Mathew *Habitable Exoplanets: Red Dwarf Systems Like TRAPPIST-1* (OCS Book 3), 2018, ASIN-B07C69B5VJ.

Anderson, Mathew *Is Anybody Out There?: An Expanded Excerpt From Our Cosmic Story* (OCS Book 2), 2017 ASIN-B071VBCX54.

Behe, Michael J. *The Edge of Evolution: The Search for the Limits of Darwinism,* 2008, ISBN-0743296222.

Campins, H.; Hargrove, K; Pinilla-Alonso, N; et al. (2010). "Water ice and organics on the surface of the asteroid 24 Themis", Nature, avail: https://www.ncbi.nlm.nih.gov/pubmed/20428164.

Catling, David C. *Astrobiology: A Very Short Introduction.* Oxford: Oxford University Press, 2013, ISBN-978-0-19-958645-5.

Cockell, Charles S. *Astrobiology: Understanding Life in the Universe,* 1st Edition, ISBN-1118913337.

Crick, Francis Life Itself: Its Origin and Nature, 1981, Simon and Schuster, ISBN-0671255622.

Crosse, Andrew, Weekes, W. H. *Abiogenesis and Life from Dirt: The Andrew Crosse Experiment,* 2015, ISBN-1943392005.

Culp, Jennifer *How We Find Other Earths: Technology and Strategies to Detect Planets Similar to Ours (Search for Other Earths),* 2016, ISBN-1499462921.

Dawkins, Richard *The Ancestor's Tale: A Pilgrimage to the Dawn of Life,* 2004, Houghton Mifflin Harcourt, ISBN–054485993.

Dick, Steven J. *Astrobiology, Discovery, and Societal Impact,* 2018, Cambridge University Press, ISBN-10-110842676X.

Dyson, Freeman *Origins of Life,* 1986, Cambridge University Press, ASIN-B001563IFC.

Gargaud, Muriel; Martin, Hervé *Young Sun, Early Earth and the Origins of Life: Lessons for Astrobiology,* 2013, ISBN-3642225519.

Goldsmith, Donald *Exoplanets: Hidden Worlds and the Quest for Extraterrestrial Life,* Harvard University Press, 2018, ASIN-B07DGJKF8N.

Goodwin, J. T. and D. G. Lynn. 2014. *Alternative Chemistries of Life – Empirical Approaches,*. ISBN: 978-0-692-24992-5.

Greenburg, Richard *Europa – The Ocean Moon: Search For An Alien Biosphere,* Springer Praxis Books, 2009, ISBN-3540224505.

Greenburg, Richard *Unmasking Europa: The Search for Life on Jupiter's Ocean Moon,* 2008, ASIN – B00E6T9UHA.

Hand, Kevin The Search for Life in the Depts of Space, 2020, ASIN-B07YYR6F4V.

Hart, Chris; Vernekar, Shubham *After Earth : The Search for Habitable Exoplanets: The Search for Habitable Exoplanets*, 2017, ASIN-B076YBDHZC.

Hays, Lindsay, ed. "NASA Astrobiology Strategy 2015" (PDF). NASA., avail:
https://nai.nasa.gov/media/medialibrary/2015/10/NASA_Astrobiology_Strategy_2015_151008.pdf

Heng, Kevin *Exoplanetary Atmospheres: Theoretical Concepts and Foundations* (Princeton Series in Astrophysics), 2017, ISBN-0691166978.

Herzing, Denise L., Johnson, Christine M. (eds) *Dolphin, Communication and Cognition: Past, Present, and Future*, 2015, MIT Press, ISBN-978-0262029674.

Hoyle, Sir Fred ; Astucia, Salvador, *Evolution from Space*, 2015, ASIN-B019UVYW2G.

Hoyle, Fred *The Intelligent Universe*, 1988, ISBN-0030700833.

Hoyle, Fred; Wickramasinghe, Chandra; Watson, John *Viruses from Space and Related Matters*, 1986, University College Cardiff Press, ISBN 978-0-906449-93-6.

Impel, Chris, et al. *Frontiers of Astrobiology*, 2012, ISBN-1107006414.

Kargel, Jeffrey S.; Kaye, Jonathan Z.; Head, James W.; Marion, Giles M.; Sassen, Roger; et al. (November 2000). "Europa's Crust and Ocean: Origin, Composition, and the Prospects for Life", Icarus, 148 .

Kenney, Karen Latchana *Exoplanets: Worlds beyond Our Solar System*, 2017, ASIN-B01N2RSA5H.

Kitchin, C. R. *Exoplanets: Finding, Exploring, and Understanding Alien Worlds*, 2012, ISBN-1461406439.

Klein, David, *Organic Chemistry*, 2nd cd, Wiley, 2013, ISBN 1118452288

Linde, Peter *The Hunt for Alien Life: A Wider Perspective (Astronomers' Universe), 2016, ISBN-3319241168.*

Machalek, Pavel, "Organic Molecules in Comets and Meteorites and life on Earth Interstellar medium Class, 2006, https://www.spaceknow.com

Meyer, Stephen C. *Signature in the Cell: DNA and the*

Evidence for Intelligent Design, ISBN-978-0061472794.
Mix, Lucas John *Life in Space: Astrobiology for Everyone,* Harvard University Press, 2009, ISBN-0674033214.

Pont, Frederic J. *Alien Skies: Planetary Atmospheres from Earth to Exoplanets, 2014, ISBN-1461485533.*

Rampelotto, P. H. "Panspermia: A Promising Field Of Research, Astrobiology Science Conference, 2010: Evolution and Life: Surviving Catastrophes and Extremes on Earth and Beyond. 20–26 April 2010. League City, Texas. Bibcode:2010LPICo1538.5224R.

Regius, Codex *Titan: Pluto's big brother: The Cassini-Huygens spacecraft and the darkest moon of Saturn,* 1st ed, 2016, ISBN-1541307453.

Regius, Codex *Ceres: Pluto's little sister: The Dawn spacecraft over the volcanoes of a dwarf planet,* 2017, ISBN-101545462933.

Rothery, David A. (Ed), Gilmour Iain (Ed), Sephton, Mark A. (Ed)
An Introduction to Astrobiology, 3rd Edition, Cambridge University Press, 2018, ISBN-978-1108430838.

Trifonov, Edward; Lane, Nick; Freeland, Stephen; Russell, Michael *Abiogenesis: How Life Began, the Origins and Search for Life,* 2011, ASIN-B005FY5ZAG.

Scharf, Caleb *The Copernicus Complex: Our Cosmic Significance in a Universe of Planets and Probabilities*, 2014, ISBN-0374129215.

Schenk, Paul M. (ed) et al *Enceladus and the Icy Moons of Saturn* (Space Science Series), 2018, ISBN-0816537070.

Silva, Stepheen *Extraterrestrial Communication Code : The Discovery, Meaning and Our Response to their Message*, 2021, ASIN-B08Y64W9WL.

Smith, Eric *The Origin and Nature of Life on Earth: The Emergence of the Fourth Geosphere*, 1st Edition, 2016, Cambridge University Press, ISBN-1107121884.

Stakem, Patrick H. *Exploration of the Gas Giants, Space Missions to Jupiter, Saturn, Uranus, and Neptune*, PRRB Publishing, 2018, ISBN-9781717814500.

Summers, Michael E. *Exoplanets: Diamond Worlds, Super Earths, Pulsar Planets, and the New Search for Life beyond Our Solar System,* 2017, Smithsonian, ISBN-1588345947.

Trifonv, Edward, Lane, Nick *Abiogenesis: How Life Began. The Origins and Search for Life*, 2011, ASIN-B005FY5ZAG.

Wall, Michael; Tate, Karen *Out There: A Scientific Guide to Alien Life, Antimatter, and Human Space Travel (For the Cosmically Curious),* 2018, ISBN-1538729377.

Watson, James The double Helix, A Personal Account of the Discovery of the Structure of DNA, 1980, ISBN-0-393-95075-1.

Webb, Stephan *If the Universe is Teeming with Aliens, where is Everybody?:Seventy-Five Solutions to the Fermi Paradox and the Problem of Extraterrestrial Life,* 2015, Springer, ASIN-B00XVTG1NC

Willis, Jon *All These Worlds Are Yours: The Scientific Search for Alien Life*, 2016 ISBN-0300208693.

Yaqoob, Tahir *Exoplanets and Alien Solar Systems,* 2011, ASIN-B005WQ0E6C.

Resources

- https://exoplanets.nasa.gov/
- https://www.nasa.gov/feature/jpl/what-in-the-world-is-an-exoplanet
- https://exoplanetarchive.ipac.caltech.edu/
- https://www.swri.org/press-release/evidence-complex-organic-molecules-enceladus
- www.planetary.org
- Space Telescope Science Institute public outreach, http://outreachoffice.stsci.edu/
- https://www.space.com/17738-exoplanets.html
- https://www.nasa.gov/tess-transiting-exoplanet-survey-satellite
- https://www.lpi.usra.edu/opag/meetings/feb2016/presentations/day-1/08-Roadmap-Ocean-Worlds-McEwen.pdf
- https://spacenews.com/ocean-worlds-discoveries-build-case-for-new-missions/
- www.planethunters.org
- http://phl.upr.edu/projects/habitable-exoplanets-catalog/exoplanet-resources
- SETI Institute – seti.org
- organic material in meteorites: - http://www.daviddarling.info/encyclopedia/C/carbchon.html.
- https://www.space.com/40819-mars-methane-organics-curiosity-rover.html
- https://www.nature.com/articles/s41586-018-0246-4.

- https://www.astrobio.net
- carlsaganinstitute.org
- https://www.jpl.nasa.gov/habex/documents/
- https://astrobiologyfuture.org/
- wikipedia, various.

If you enjoyed this book, you might also be interested in some of these.

Stakem, Patrick H. *16-bit Microprocessors, History and Architecture*, 2013 PRRB Publishing, ISBN-1520210922.

Stakem, Patrick H. *4- and 8-bit Microprocessors, Architecture and History*, 2013, PRRB Publishing, ISBN-152021572X,

Stakem, Patrick H. *Apollo's Computers*, 2014, PRRB Publishing, ISBN-1520215800.

Stakem, Patrick H. *The Architecture and Applications of the ARM Microprocessors*, 2013, PRRB Publishing, ISBN-1520215843.

Stakem, Patrick H. *Earth Rovers: for Exploration and Environmental Monitoring*, 2014, PRRB Publishing, ISBN-152021586X.

Stakem, Patrick H. *Embedded Computer Systems, Volume 1, Introduction and Architecture*, 2013, PRRB Publishing, ISBN-1520215959.

Stakem, Patrick H. *The History of Spacecraft Computers from the V-2 to the Space Station*, 2013, PRRB Publishing, ISBN-1520216181.

Stakem, Patrick H. *Floating Point Computation*, 2013,

PRRB Publishing, ISBN-152021619X.

Stakem, Patrick H. *Architecture of Massively Parallel Microprocessor Systems*, 2011, PRRB Publishing, ISBN-1520250061.

Stakem, Patrick H. *Multicore Computer Architecture,* 2014, PRRB Publishing, ISBN-1520241372.

Stakem, Patrick H. *Personal Robots*, 2014, PRRB Publishing, ISBN-1520216254.

Stakem, Patrick H. *RISC Microprocessors, History and Overview,* 2013, PRRB Publishing, ISBN-1520216289.

Stakem, Patrick H. *Robots and Telerobots in Space Applications*, 2011, PRRB Publishing, ISBN-1520210361.

Stakem, Patrick H. *The Saturn Rocket and the Pegasus Missions, 1965,* 2013, PRRB Publishing, ISBN-1520209916.

Stakem, Patrick H. *Visiting the NASA Centers, and Locations of Historic Rockets & Spacecraft,* 2017, PRRB Publishing, ISBN-1549651205.

Stakem, Patrick H. *Microprocessors in Space*, 2011, PRRB Publishing, ISBN-1520216343.

Stakem, Patrick H. Computer *Virtualization and the*

Cloud, 2013, PRRB Publishing, ISBN-152021636X.

Stakem, Patrick H. *What's the Worst That Could Happen? Bad Assumptions, Ignorance, Failures and Screw-ups in Engineering Projects, 2014,* PRRB Publishing, ISBN-1520207166.

Stakem, Patrick H. *Computer Architecture & Programming of the Intel x86 Family, 2013,* PRRB Publishing, ISBN-1520263724.

Stakem, Patrick H. *The Hardware and Software Architecture of the Transputer,* 2011,PRRB Publishing, ISBN-152020681X.

Stakem, Patrick H. *Mainframes, Computing on Big Iron,* 2015, PRRB Publishing, ISBN- 1520216459.

Stakem, Patrick H. *Spacecraft Control Centers,* 2015, PRRB Publishing, ISBN-1520200617.

Stakem, Patrick H. *Embedded in Space,* 2015, PRRB Publishing, ISBN-1520215916.

Stakem, Patrick H. *A Practitioner's Guide to RISC Microprocessor Architecture,* Wiley-Interscience, 1996, ISBN-0471130184.

Stakem, Patrick H. *Cubesat Engineering,* PRRB Publishing, 2017, ISBN-1520754019.

Stakem, Patrick H. *Cubesat Operations*, PRRB Publishing, 2017, ISBN-152076717X.

Stakem, Patrick H. *Interplanetary Cubesats*, PRRB Publishing, 2017, ISBN-1520766173 .

Stakem, Patrick H. Cubesat Constellations, Clusters, and Swarms, Stakem, PRRB Publishing, 2017, ISBN-1520767544.

Stakem, Patrick H. *Graphics Processing Units, an overview*, 2017, PRRB Publishing, ISBN-1520879695.

Stakem, Patrick H. *Intel Embedded and the Arduino-101, 2017,* PRRB Publishing, ISBN-1520879296.

Stakem, Patrick H. *Orbital Debris, the problem and the mitigation*, 2018, PRRB Publishing, ISBN-*1980466483.*

Stakem, Patrick H. *Manufacturing in Space*, 2018, PRRB Publishing, ISBN-1977076041.

Stakem, Patrick H. *NASA's Ships and Planes*, 2018, PRRB Publishing, ISBN-1977076823.

Stakem, Patrick H. *Space Tourism*, 2018, PRRB Publishing, ISBN-1977073506.

Stakem, Patrick H. *STEM – Data Storage and Communications*, 2018, PRRB Publishing, ISBN-1977073115.

Stakem, Patrick H. *In-Space Robotic Repair and Servicing*, 2018, PRRB Publishing, ISBN-1980478236.

Stakem, Patrick H. *Introducing Weather in the pre-K to 12 Curricula, A Resource Guide for Educators*, 2017, PRRB Publishing, ISBN-1980638241.

Stakem, Patrick H. *Introducing Astronomy in the pre-K to 12 Curricula, A Resource Guide for Educators*, 2017, PRRB Publishing, ISBN-198104065X.

Also available in a Brazilian Portuguese edition, ISBN-1983106127.

Stakem, Patrick H. *Deep Space Gateways, the Moon and Beyond*, 2017, PRRB Publishing, ISBN-1973465701.

Stakem, Patrick H. *Exploration of the Gas Giants, Space Missions to Jupiter, Saturn, Uranus, and Neptune*, PRRB Publishing, 2018, ISBN-9781717814500.

Stakem, Patrick H. *Crewed Spacecraft*, 2017, PRRB Publishing, ISBN-1549992406.

Stakem, Patrick H. *Rocketplanes to Space*, 2017, PRRB Publishing, ISBN-1549992589.

Stakem, Patrick H. *Crewed Space Stations,* 2017, PRRB Publishing, ISBN-1549992228.

Stakem, Patrick H. *Enviro-bots for STEM: Using Robotics in the pre-K to 12 Curricula, A Resource Guide for Educators,* 2017, PRRB Publishing, ISBN-1549656619.

Stakem, Patrick H. *STEM-Sat, Using Cubesats in the pre-K to 12 Curricula, A Resource Guide for Educators*, 2017, ISBN-1549656376.

Stakem, Patrick H. *Lunar Orbital Platform-Gateway*, 2018, PRRB Publishing, ISBN-1980498628.

Stakem, Patrick H. *Embedded GPU's*, 2018, PRRB Publishing, ISBN- 1980476497.

Stakem, Patrick H. *Mobile Cloud Robotics*, 2018, PRRB Publishing, ISBN- 1980488088.

Stakem, Patrick H. *Extreme Environment Embedded Systems,* 2017, PRRB Publishing, ISBN-1520215967.

Stakem, Patrick H. *What's the Worst, Volume-2*, 2018, ISBN-1981005579.

Stakem, Patrick H., *Spaceports*, 2018, ISBN-1981022287.

Stakem, Patrick H., *Space Launch Vehicles*, 2018, ISBN-1983071773.

Stakem, Patrick H. *Mars*, 2018, ISBN-1983116902.

Stakem, Patrick H. *X-86, 40th Anniversary ed*, 2018, ISBN-1983189405.

Stakem, Patrick H. *Lunar Orbital Platform-Gateway*, 2018, PRRB Publishing, ISBN-1980498628.

Stakem, Patrick H. *Space Weather*, 2018, ISBN-1723904023.

Stakem, Patrick H. *STEM-Engineering Process*, 2017, ISBN-1983196517.

Stakem, Patrick H. *Space Telescopes*, 2018, PRRB Publishing, ISBN-1728728568.

Stakem, Patrick H. *Exoplanets*, 2018, PRRB Publishing, ISBN-9781731385055.

Stakem, Patrick H. *Planetary Defense*, 2018, PRRB Publishing, ISBN-9781731001207.

Patrick H. Stakem *Exploration of the Asteroid Belt*, 2018, PRRB Publishing, ISBN-1731049846.

Patrick H. Stakem *Terraforming*, 2018, PRRB Publishing, ISBN-1790308100.

Patrick H. Stakem, *Martian Railroad*, 2019, PRRB Publishing, ISBN-1794488243.

Patrick H. Stakem, *Exoplanets,* 2019, PRRB Publishing, ISBN-1731385056.

Patrick H. Stakem, *Exploiting the Moon,* 2019, PRRB Publishing, ISBN-1091057850.

Patrick H. Stakem, *RISC-V, an Open Source Solution for Space Flight Computers,* 2019, PRRB Publishing, ISBN-1796434388.

Patrick H. Stakem, *Arm in Space,* 2019, PRRB Publishing, ISBN-9781099789137.

Patrick H. Stakem, *Extraterrestrial Life,* 2019, PRRB Publishing, ISBN-978-1072072188.

Patrick H. Stakem, *Space Command,* 2019, PRRB Publishing, ISBN-978-1693005398.

CubeRovers, A Synergy of Technologys, 2020, PRRB Publishing, ISBN-979-8651773138.

Robotic Exploration of the Icy moons of the Gas Giants. 2020, PRRB Publishing, ISBN- 979-8621431006

Hacking Cubesats, 2020, PRRB Publishing, ISBN-979-8623458964.

History & Future of Cubesats, PRRB Publishing, ISBN-979-8649179386.

Hacking Cubesats, Cybersecurity in Space, 2020, PRRB Publishing, ISBN-979-8623458964.

Powerships, Powerbarges, Floating Wind Farms: electricity when and where you need it, 2021, PRRB Publishing, ISBN-979-8716199477.

Hospital Ships, Trains, and Aircraft, 2020, PRRB Publishing, ISBN-979-8642944349.

2020/2021 Releases

CubeRovers, a Synergy of Technologys, 2020, ISBN-979-8651773138

Exploration of Lunar & Martian Lava Tubes by Cube-X, ISBN-979-8621435325.

Robotic Exploration of the Icy moons of the Gas Giants, ISBN- 979-8621431006.

History & Future of Cubesats, ISBN-978-1986536356.

Robotic Exploration of the Icy Moons of the Ice Giants, by Swarms of Cubesats, ISBN-979-8621431006.

Swarm Robotics, ISBN-979-8534505948.

Introduction to Electric Power Systems, ISBN-979-8519208727.

Centros de Control: Operaciones en Satélites del Estándar CubeSat (Spanish Edition), 2021, ISBN-979-8510113068.

Exploration of Venus, 2022, ISBN-979-8484416110.

Patrick H. Stakem, *The Search for Extraterrestial Life,* 2019, PRRB Publishing, ISBN-1072072181.

The Artemis Missions, Return to the Moon, and on to Mars, 2021, ISBN-979-8490532361.

James Webb Space Telescope. A New Era in Astronomy, 2021, ISBN-979-8773857969.

Riverine Ironclads, Submarines, and Aircraft Carriers of the American Civil War, 2019, ISBN- 978-1089379287.

www.ingramcontent.com/pod-product-compliance
Lightning Source LLC
Chambersburg PA
CBHW020929180526
45163CB00007B/2949